Soil Science for Regenerative Agriculture

An In-Depth Guide to No-Till Cultivation, Composting, And Natural Farming

NORBERTO M. WRAY

Table of Contents

Chapter 1. Introduction

The development of regenerative agriculture is supported by the knowledge of soil science and the technical capacity to implement many soil-ameliorating practices, which are designed to enhance the soil's ability to sustain and nourish a productive plant community. Indeed, according to best estimates, over 40% of the soils worldwide have been negatively impacted by human activities, to the extent that they are eroding and releasing carbon and nutrients faster than they are replaced. Currently, the rate of topsoil degradation is approximately 10 to 40 times the amount of soil forming under natural conditions, and while different agricultural practices can mitigate the problem to some extent, current trends predict that rapidly expanding population pressures will only make the problem worse in the future.

In recent years, the concept of "regenerative agriculture" has gained momentum, as it became apparent that the current approaches to crop and livestock production are limited in their ability to meet global food and environmental challenges. We now understand that to curb climate change, we need to not only reduce carbon, nitrous, and methane emissions at their sources but also sequester more carbon into the soils and plants.

1.1 Unlocking the Future of Sustainable Farming

There is a cyclical relationship between plant and soil health. The soil provides the plant with the water, nutrients, and root environment necessary for the plant to grow. Then the plant will deliver nutrients, organic matter, and carbon-based exudates to the soil. These exudates, in particular, play an important role in shaping the soil ecosystem, which consequently influences the health of the plant through improved nutrient availability, protection against pathogens, and enhanced tolerance to pests and abiotic stress.

In the broader agricultural system, the choices made by farmers and downstream actors will shape these plant-soil interactions. These choices will likely affect soil health and thus the stress tolerance of the plant. Intensive farming practices and the use of high levels of artificial chemical inputs will often lead to soil degradation. Impacts of soil degradation include lower levels of organic matter, which can have implications for soil structure and soil biology. Indeed, soil biology will only be supported by the levels of organic matter remaining in the soil. The long-term overuse of artificial chemical inputs can then potentially lead to the development of soil fertility problems, as well as create a lower-level stress environment for pests and diseases to exploit.

1.2. The Transformative Power of Regenerative Agriculture

The nature of hydrolyzing and microorganisms establishes healthy soil and provides the foundation for regenerative agriculture. Healthy soil sustains crop and forage production. Very high levels of microbial diversity and active faunal populations are essential. Most important is arbuscular mycorrhizal with roots of most plants and fungal-dominated food webs due to the demonstrated ability to produce specific organic compounds that are the precursors of crumb structure called glomalin. The most powerful nutrient released, in addition to nutrient-rich bliss, is the arbuscular mycorrhizal glomalin that helps reestablish beneficial groups, especially sugars and related carbohydrate plant symbioses, using Sesquicarbonate weathering as an energy and carbon source through biomass that includes arbuscular.

Regenerated soil plays a key role in regenerating the water cycle, and by doing this, effectively makes the climate on earth responsible for changing deserts, depleting aquifers, increasing rainfall, increasing flooding, and subsequent regulatory changes. As incredible as it seems, little life is known about the microbial and microfaunal structures that are too adapted to call soil did kak, and no exceptions. Reviving and maintaining soil health is primarily addressed by reducing soil destruction in bases and improving the soil's lack of water precipitation that slows down surface runoff, recharge infiltration, and soil degradation. Before soil health impedes production, benefits the ultramacronutrientated processes.

1.3. Practical Benefits for Farmers and Gardeners

For the farmer or gardener, the other benefit of soil containing large macro-pores and large soil aggregates with good soil structure, large organic matter content, predominantly carbon, that is most important of all, is **"The Denitrification Filtration of Groundwater"** special feature. This states that the beneficial relationships between plants, fungi, and soil are important not just in sourcing water and nutrients for the plant, but also in protecting groundwater quality. These relationships contain a suite of services critical to reducing the movement of nitrates to groundwater through a process known as denitrification. As nitrates pass through the soil with the water flow, they are removed from the water column as they pass through zones of soil containing good supplies of carbon, and a large range of fungi and bacteria rapidly break the nitrate molecule down and use it as an energy source, ultimately releasing harmless nitrogen gas to the air.

By increasing the organic matter, predominantly carbon, in the soil and by creating good soil structure, large macro-pores, and large water-stable aggregates, the movement of water and gases in the soil profile will be improved. This should allow water to enter the soil more readily and travel through the soil more easily, giving improved water infiltration. Improving soil structure and creating aggregates will also help to reduce and eliminate soil erosion through surface water runoff. The more water-stable organic matter in the soil, derived from the decay of roots, root exudates, mycorrhizal exudates, stress metabolites shed by plant roots, and root hairs during the process of converting carbon into useful carbohydrates, will help to bind the soil and aggregates together. This will reduce the risk of soil erosion by water and resulting soil loss.

1.4. How This Guide Will Revolutionize Your Approach

To understand soil, one has to go deeper into the science of soil science. Soil science is the study of soil as a natural resource on the surface of the earth, including its formation, properties, hydrology, biology, and use. It is closely related to other diverse disciplines: chemistry, physics, engineering, ecology, and even health! This book can assist any prospective or practicing geologist, biologist, hydrologist, agronomist, and others. However, it has been formatted specially for the use of farmers, gardeners, engineers, public servants, and other specialists who work with the land, who need to understand and, in turn, describe the soils they encounter and work with.

By emphasizing natural soil classifications and the physical-chemical properties that control soil behavior, productivity, and fertility, sections on soils under special kinds of use can be brief. Most importantly, the suggested system provides emphasis to young geologists and scholars preparing for professions in which they will be working with the very thin coating of little understood and relatively young soil on the terrestrial part of the Earth's surface. As every generation has to start again in building upon the soil-related knowledge of their predecessors, they must begin with a scheme that reflects current views. By the way, the system also is designed to fit the use of the engineer, completing an interdisciplinary approach to teaching and investigating the soil.

Part I: The Basics of Soil Science

It is no longer a secret that people and their agricultural activities have a greater impact on the Earth's ecosystem than any other known force in nature. However, what's promising is that this can change when people focus on nature's basic systems and team up with them in a regenerative manner. The heart and soul of regenerative agriculture are the basic resiliency that comes from healthy, vibrant soil. In essence, regenerative agriculture begins with the soil and the foundational understanding of the basic principles of soil creation, structure, nutrients, and biological communities that comprise both this structure and nutrients. In this article and the ones that follow, we will explore the various aspects of the soil and living systems within.

Soil Science reached its present status through a very slow process. Fundamental facts are becoming clear now; hence the progress of soil science has been quick during the last quarter of a century. But the scientific synthetic view has not been published till today, not even till 30 percent of it is learned. It only means that the science cannot make progress by getting a set of particular facts and principles correctly and clearing the immediate background. The science cannot be expressed as a finished product at any particular time.

Chapter 2: Understanding Soil Science

The medium in which plants grow is the soil profile, which consists mostly of mineral particles and a small fraction of plant roots, decaying organic matter, living organisms, and water and gas-filled pores. Near roots - the rhizosphere - the soil profile plays a vital role as the plant develops, influenced in turn by water and nutrient availability, and other soil characteristics. Management strategies for soil, crop, and animal interactions arise from a sound understanding of the nature and characteristics of biologically active and mineral contents. Such an understanding involves comprehension of the interactions between geology, topography, vegetation, climate, and how these have evolved to create different categories of soil.

This basic knowledge is entrenched in soil science. The considerable body of data and information that form the foundation for cropping and livestock procedures in diverse regions took hundreds, and sometimes thousands, of years to develop. When such knowledge is ignored, it is long forgotten or abandoned, often to the detriment of economic return, the terrestrial environment, and natural resources in general, and not just soil. Crown practitioners of regenerative techniques recognize the need to work with this suite of resources, prompted by the conservation and restoration of ecosystems for their benefits to biodiversity, together with the necessity of ensuring long-term sustainability in food, feed, fiber, and biofuel production. Sandier soil supports poor, while the gravely soil, inherited from an aerosol, rapidly loses fertility. Large areas now suffer severe land resulting from inadequate management in the past, along with a history of high levels of erosion or compaction.

2.1. Fundamentals of Soil Composition and Health

Historically, agriculture has been about the removal of life from the soil. Conventional agriculture is not only inefficient, but it is also liquidating the very resource on which it depends. The essence of organic farming and biologically based regenerative agriculture is to work in concert with the natural ecology instead of suppressing and overpowering it. Healthy soil is created and sustained by re-establishing ecological balance, and this key lies in the microbes thriving in the soil. There is no substitute for the complex array of functional activity carried out by beneficial soil microorganisms. Soil is an ecosystem under our feet composed of different components like the sand, silt, and clay particles, organic matter, minerals, water, air, and living things (plants, insects, worms, microbes, etc.). There are more living things in a single teaspoon of soil than there are people on earth. Soils sustain plant and animal life and act as a biological filter to clean air and water.

Soils are where the atmosphere and the earth's lithosphere interact, and they are as intricate and complex as they are diverse. The earth's lithosphere is the 'crust' of the earth upon which we live; it has many layers, and soil scientists group them into categories mainly based on their shapes, sizes, and chemical compositions. The earth's lithosphere is really a skin-like shell, and although it appears thick, the living layer of the earth, where the lithosphere and the lively atmosphere interact, is only about 10 feet deep. It is estimated that it takes 500 years on average to create just an inch of soil. Building healthy soil is about building soil that is full of life. Over the years, soil has been seen as an inert growing medium that anchors plants and acts as a reservoir of mineral elements that plants need. Soil was often ignored and treated as an article in an equation of low-cost production. Soil is a habitat, a living system that needs to be treated in a complex, ecologically sustainable way over the long term.

2.2. Ancient Techniques and Modern Insights

Soil depth, structure, water holding capacity, nutrient value, and other soil properties are all affected by the life thriving in the soil. Traditional societies knew this without the data available today and incorporated numerous practices designed to care for the land and make it more productive. From a confluence of folk wisdom and scientific discovery, we now know that both the traditional societies of the past and modern regenerative agriculture practices are effective because they improve soil health. As we learn more with each passing year, the reasons for these successes are also a reflection of the principles of ecology and biology about how organisms that exist in the soil and the land they live in can help boost the capacity of the land to produce food and other goods.

The practices and treatments involved in regenerative agriculture all build on several fundamental principles of soil science. The possibilities for regenerating land and the primary building blocks of regenerative agriculture arise from interactions of organic carbon, chemical products, minerals, and moisture. The most important of these building blocks are the mineral and organic components of the soil. The inorganic or mineral part of the soil can represent a very broad range of chemical compounds. These are typically a combination of oxygen, silicon, aluminum, iron, and other primary elements. Organic matter is what makes soil able to provide ecosystem functions, both for agricultural production and when we allow for the preservation or regeneration of natural habitats.

2.3. Practical Applications of Soil Science

The active carbon and basal respiration tests give a lot of information about carbon cycling in the soil environment. These tests can be done quickly and give relative comparisons of soil biological activity. Measuring soil organic matter quantitatively involves measuring its concentration. The Wet Combustion Method burns the sample in a 360°C oven and measures the carbon dioxide and water which are produced. Because the reactions are conducted in liquid solutions and involve complex chemical reactions to be controlled, numerous errors can occur. (For example, contamination of the samples with particles from the air or reagent vessel, errors in measurements or amounts added by the analyst, problems associated with volatile fiber method).

Test speed, cost, equipment needed, difficulty, safety considerations, and need for special expertise are used to determine how easy or difficult a test is to conduct. Soil analysis is itself a cheap and dirty business. Abundant soil organic matter ensures superior soil structure and more biomass production. However, merely adding organic matter does not always achieve these results; rather, the increased organic matter addition needs to stimulate the soil biotic community to thrive, multiply, and function. Soil structure and local environmental conditions set the stage for these results to occur.

The success of any crop production prescribed burning depends on good techniques of burning. Use of a firebreak in burning causes less burning to go out of control. Fertilizer requirement for high yield is subject to the amount of erosion and the type of crop planted. Not all areas require the same amount of fertilizers. Gradual increment as soil fertility decreases should be adopted. For best effect, fertilization should be done at the right time. Use of the right fertilizer formulation should also be considered. Foliar fertilization is especially beneficial to crops developing yellowish leaves. Higher fertilizer use to improve soil fertility may decrease the number of days necessary for growing crops to mature.

Chapter 3: The Role of Soil in Ecosystems

Soil performs several crucial ecosystem functions which are the criteria for an ecosystem rather than a human-centric concept of good soil. These include sequestration of carbon, breakdown of organic matter (primarily to produce carbon dioxide and water) in respiration and other metabolic processes, support and nutrition of vegetation, support of animal life, and influencing climate, particularly local temperature and humidity. Soil sequesters carbon in the biota, and this adds to the productivity of the system as detritivores produce greenhouse gases. Soil organic carbon also contributes to high levels of fertility. Some stores of carbon are highly adapted to their function, even if this increases their longevity, such as the structure within peds. Traditionally, carbon itself, particularly long molecules, is considered to be recalcitrant, so the products of food chains in mineral soil contain even more recalcitrant materials.

All products of the food chain become part of the soil matrix, and turnover initially relies on them accumulating. Partly, we save the corpses of plant roots that occurred as the plants exude secretions and because the tissues lignify as plants die. Dead microbial biomass has a similar composition. Other aspects of respiration will release materials through roots and soil organisms. Individual residues may be lost due to mobility. Increased soil carbon stocks coincide with the use of conservation agriculture. Establishing grass in the rotation and leaving long-lying residues are the most effective, and also higher levels of soil microbial activity. Soils stock more nitrous oxide in tiles. Raised water tables, particularly as GHG production takes place within a soil micro-site and there is no excess moisture. It's also seen settling carbon in the biota reducing carbon loss to make the emission of nitrous oxide more important. Increased retention of natural moisture levels can be very beneficial to plant productivity; adding organic matter can produce complex aggregates that water held near peds of plowing, while the aggregate hanging from the plan hinges around their fully expanded structures.

3.1. Importance of Soil Biodiversity

Soil is teeming with life. The most abundant form of life on Earth resides in the soil, where millions of organisms live in only a handful of soil. These organisms are essential components of the ecosystem and provide vital ecosystem services, including nutrient cycling, pest control, disease suppression, and carbon sequestration. As a direct product of the organisms living in the soil, soil structure and fertility are largely determined by the biodiversity of the soil. Higher levels of biodiversity above ground, such as in the fields of large-scale agriculture, are directly related to higher levels of biodiversity in the soil.

When it comes to rebuilding soil biodiversity in the fields of large-scale agriculture, two main strategies are commonly employed. Additive strategies include practices that add organic matter, such as compost, cover crops, and straw, to the soil. However, soils that are deprived of organic matter from years of intensive tillage cannot always support high levels of life even when they are provisioned with materials such as this. These materials can disappear through decomposition, alongside some consumers and their manure. Pioneering soil management practices such as agroecology, which mimic natural systems in a resource-efficient manner through "fostering synergy with natural ecosystem processes to optimize the use of ecological nutrient flows," are paramount for fostering the regeneration of soil ecosystems through an increase in soil life. Multiple studies have demonstrated that reducing disturbance and implementing diversified crop rotations increases soil carbon, nutrients, and the abundance of soil life within a few years of transitioning to agroecological management.

3.2. Impact of Soil Quality on Plant Health

Plants need much more than the elements in fertilizer. This observation is at the heart of shifting agriculture from a mineral fertilization and pest-management model to a holistic plant health and nutrient-holding capacity strategy that relies on soil biology, organic matter, mineral nutrition, and soil physical properties. A functional plant is a luxury item for the plant itself, and we need to make sure that the plant reaches its full potential. The parts of a plant all work together and form an integrated unit. When one part does not function perfectly, the associated parts have to take over its role. Only when the machinery of the individual parts has its full potential, the machine of the whole plant can work with its full potential. Postponing remedial action often means that the plant's capacity for recovery is reduced. When optimizing plant health is the focus of the grower, there is less need for problem-solving.

The initial goal for soil management is to select practices to achieve field-requirement specifications for the physical, mineral, organic matter, and biological components of farm soil. These critical component specifications have an integrated function in promoting plant health while minimizing disturbance of nature's order, and the resulting synergy can optimize the system and result in greater reliability and economic yield. The resultant system feedback effect, by creating a soil that is conducive to high-yielding, healthy, sustainable crops, inevitably will:

1. delay systems failure;
2. reduce the potential for plant pests or impeding the efficacy of crops that are intended to become or to remain pests;
3. Lessen or mitigate the virtual certainty of a pesticide treadmill, or worse, if the nutritional requirements of the plant are not met.

Moreover, our cost of production could be reduced at the same time we improve our soils.

3.3. Soil's Role in Climate Regulation and Water Cycles

Atmospheric systems on many planets may maintain stable surface temperatures through energy exchange with space or internal thermal processes, but Earth's atmospheric system takes a less direct route. The global climate is activated by solar radiation but is moderated (to some extent) by the availability and distribution of energy provided by water and earth materials organized by life processes. Climate is, in turn, a major factor in regional distributions of life systems, landforms, frozen water, and solid and liquid water bodies, and strongly influences how actually-navigable Earth's surface is by human societies and which regions are populated by what types of these societies. The holon of ecosystems climatically regulates energy release. Earth's service to the solar system climate system is bi-directional. Ecosystems are sometimes stressed and even destroyed when temperatures are too high or too low, or when precipitation is either too frequent or too rare, in spatial and temporal contexts that are predictable based on the combination of regional energy release and water dispersion patterns.

Soil water is quite important to the ecosystems that are described in much of this volume as water-limited. Root systems are not designed to absorb more water than the soil can make available. Managed living landscapes have moisture relationships that relate to soil surface evaporation and runoff processes, but unmanaged native landscapes develop substantial subsurface reserves of water which root systems utilize to survive annual periods of drought. Productive vast natural ecosystems, from rainforests to deserts, express annual total plant growth processes that are constructed from mass-matching components. The lives and life cycles of their various organism categories have the elements of the biological pulsing chapters.

Part II: Regenerative Practices

Many approaches and practices contribute to achieving regenerative agriculture: practices that increase the organic matter content of soils, improve soil nutrient and water cycling, enhance resiliency to droughts and floods, improve productivity, and revitalize rural communities. The case studies make clear that revitalizing the soil with regenerative agriculture strategies can generate powerful positive benefits not only for productivity but also for the diversity of crops grown and markets served, for waste reclamation, energy generation, and environmental management, and for the economic, social, and environmental health, livelihoods and security, and societal vision and potential of farmed areas and degraded lands undergoing restoration.

Among the key practices are increasing organic matter content in soil via cover crops and reduced tillage; maximizing biodiversity with diverse and synergistic crop and animal systems, cover crops, and perennials in agricultural systems; maintaining a living root; minimizing the reliance on high energy, synthetic inputs; and integrating crop development and its consequences with natural and social environments. This set of practices, while not comprehensive, is particularly well suited to revitalize farmers' operational health, increase the productivity and resilience of large and small-scale food systems, regenerate soils, revive rural communities, and adapt to a changing climate. Regenerative agriculture can contribute to sustainable food and agriculture and positive economy-wide outcomes if practiced at scale and over time by established producers, new entrants, and communities around the world. Regeneration can be achieved on large commercial farms and small-scale farms, indoors and outdoors, and in every country. *This chapter is devoted to regenerative practices used in crop and animal production, perennials, and waste and associated management practices.*

Chapter 4. Reviving the Earth with Regenerative Agriculture

Regenerative farming principles reconceptualize agriculture as ecosystem stewardship. They focus on healing the earth through honoring and rejuvenating all life. They depend on a deep systemic understanding of the natural science and artistic expression of farming. These principles center on increasing photosynthesis while minimizing soil disturbance and enhancing diversity, both above and below ground. Reviving and understanding healthy water, mineral, and soil cycles is their purpose. Like all important work to save the commons, the scientific support for regenerative practices is strong, and they are not inferior in profits in the hands of skilled practitioners. The fact that these regenerative principles mimic natural processes is a testament to nature's perfect wisdom. By paralleling natural systems, but with human intent, regenerative agriculture leverages the extraordinary powers of life.

Veteran farmers below in the Midwest have found, refined, and promoted these new ways to frame profitable farming in harmony with the earth. We have become educated in the process of these same principles in growing healthy businesses. Yes, regenerative agricultural practices require and regenerate economic and ecological virtues simultaneously. Please join us for this examination and celebration of the infinite power of intrinsic Earth healing principles when our intent is aligned with nature's design.

4.1. Core Principles and Tangible Benefits

There are several interconnected core principles of regenerative agriculture that work in harmony to enable the benefits described here. The "soil health" principle emphasizes the maintenance or enhancement of a living system, the soil microbiome, which often dominates microbial life on Earth. The "ecological and water health" principle makes sure that ecosystem balance is neither affected by nor contributes to, the destabilization of climate and hydrologic cycles. The "animal welfare" principle heightens care of sentient beings that are directly connected to, and affected by, the farming process. Utilization of the collaboration and knowledge of the local community as well as local genetics for plants, animals, and soil microbiomes, to fit into or establish balance with local ecology, to minimize the requirement for newly imported genetics, is explicitly or implicitly involved in these principles.

The first benefit is the sequestration of the excess CO_2 released during the Industrial Revolution. Topsoil is the main Earth system able to capture the excess carbon, still in the atmosphere. The biological process occurring during soil building contains elements that are capable of snatching and sequestering CO_2 from the air on a 'carbon negative' basis. Further, increased photosynthetic activity by plants to produce high carbohydrate root exudates can sequester carbon dioxide at a near net-zero cost. This carbon capture can be tremendously enhanced by transitions from annual monocultures to perennial polycultures. Biologically active soil employs approximately the carbon dioxide domestically to feed and reproduce members of the soil microbiome, reduce the greenhouse effect, and reconnect a carbon limitation with a newly established energy limitation. Soils enriched by living biology are held together by aggregating ecosystems built of carbohydrates. The initial input of carbohydrates is obtained from plant biomass, while the whole process is eventually driven by root exudates.

4.2. Strategies for Implementation

The introduction of regenerative agriculture into farming lands involves many steps. As with climate change, there are numerous paths to the same goal of revitalizing our soil. The evidence base is the first step in the transition to regenerative agriculture. Proving its contribution to exacerbating climate change, erosion, and loss of biodiversity among other problems is a key step in its acceptance and perhaps inclusion in agri-environmental schemes. But acceptance and adoption by all members of a community of practice will be socially mediated, depending on farmers' personalities, contacts, relationships, the level of trust in methods (technological hand, reassuring hand, or political option), and also the role of knowledge and the institutional environment.

Setting up institutional conditions that encourage experimentation, farmer learning, and the sharing of experiences are also important. The introduction of monitoring and certification could play a role in terms of creating credibility at the world level (as is the case with organic agriculture). Regenerative agriculture implies moral issues (maintaining the land for future generations), utilitarian goals (higher profitability), and even ideological dimensions linked to the place of science, the role of experts, the balance between individual action and collective action, and other such matters; it has implications for the global food system and its relationships with territory, landscape, the right to food, food sovereignty, and local food infrastructures.

Chapter 5: No-Till Gardening: Techniques, Benefits, and Applications

Vegetables were grown in three soil management systems: compost-crimped rye mulch, living hairy vetch mulch, and bare soil without living cover. The living mulch treatments significantly reduced soil erosion during the pre-crop and post-crop periods compared to the compost-crimped rye and bare soil treatments. Infiltration rates were similar in the living mulch and compost-crimped rye treatments and significantly lower in the bare soil treatment. Growers that want to improve soil physical and biological health, while reducing the costs of fertility inputs and irrigation, would benefit from growing transplants with living mulch.

The potential in agricultural soil is considerably bigger. Only to keep the soil biology and organisms happy, you need a lot of plants and organic matter in the system. In the case of commercial crops, it might be more appropriate to leave the crop residues on the surface, protecting the soil from the sun and rainfalls, keeping the humidity and organic life, and avoiding using chemical treatments. We can find hundreds of edible crops for the no-till gardening setup that do not require disturbing the organic life in the first place. Eating the fruits and leaves is the most ethical way of harvesting. You do not need to disturb the underground part of veggies to eat healthy.

5.1. Benefits of No-Till Methods for Soil Health

No-till cultivation is a method used in the cultivation of crops, which is called no-till. Zero-till cultivation refers to where crops are grown with minimal soil disturbance by soil cultivation or heavy tillage practices. Crops are planted directly into undisturbed soil with minimal disturbance to the planted row, with crop residues as the only source of external soil cover. This minimal soil disturbance associated with no-till reduces erosion. Crop residues that cover the soil help restore soil organic matter. When no-till is combined with a crop rotation, soil physical properties can be rebuilt in less than 10 years.

No-till is an umbrella under which several philosophies of farming are growing. For some, it is a way to mitigate soil erosion, and reduce equipment and labor expenses, with savings of time and fuel. For others, it is more of a holistic approach to agriculture which can potentially lead to improved biological and social sustainability. No-till has the potential to combat soil erosion using water or wind. Two main factors responsible for erosion are water and wind. There are also biological characteristics that can effectively help prevent erosion if properly managed. These include structural stability, resistance to disintegration due to stress (chemical, thermal, UV, and moisture) and time, permeability and ease of water penetration, and resistance to surface water. Small phytochemicals that include lipids, organic targets, hormones, and influence metabolic activity. Most directly bound cations that contribute to soil structure and function will directly affect mold, adhesion, and cohesion.

5.1.1. Preservation of Soil Structure and Microbial Activity

Soil preservation of the structural properties, including soil structure and porosity, is important for the maintenance of soil quality. This is because soil physical properties determine the availability of essential resources, such as water and nutrients.

Soil structure consists of the spatial arrangement of primary soil particles into groups called peds or aggregates. Soil structure is the basic unit of soil organization, the building block of soil, and is situated at the edge of the passages providing pore space. Pores are the empty spaces in soil and are important for the passage and storage of water, air, solutes, and microorganisms. Two types of pores are distinguished - micropores > 30 mm in diameter, due to soil structural porosity, meso-pores 30-75 mm in diameter, due to both structural porosity and also biological activity, and macro-pores < 0.5 inches in height due to biological activity.

The soil porosity depends on the soil texture, structure, and the distribution of mineral particles. The soil porosity is formed from the original sand, silt, and clay particles and the induced pores created by soil biota. A good soil structure has approximately 50% of the soil volume as pores that are relatively evenly distributed. This is beneficial to soil health and is enhanced by the presence of bacterial MPs, bacterial hydrogels, and different soil structure modulators. Eventually, larger pore spaces can be filled with roots, macro, and meso pores based on the activity of earthworms and numerous other soil organisms.

5.2. Step-by-Step Guide to No-Till Gardening

Start with a few weeks of using soil surface coverage, preferably as sheet mulch, to suppress all weeds. Cut or remove large plants and leave what is on the soil surface. Incorporate leaves and small debris by hand and with a rake by pushing them into the top 1" of the soil with a hand fork if necessary, then planting with a hand trowel. Spread used potting mix at 25% by volume where seeds are to be sown. Bare soil could be needed for large-seeded crops like cabbage, cucumbers, and peppers; no seed treatment, no possibilities of direct-seeding crops; and cottony seedling disease. Except for those few conditions, you can create a no-till garden.

At the end of the growing season, cut the plants at the soil level, leaving the roots. The following spring, plant directly into the roots. Sheet mulching will be needed much less frequently than for typical no-till gardening.

5.2.1 Planting and Maintenance Techniques

Planting and maintenance are among the heaviest tasks of a direct seeding establishment because to ensure abandonment of slash and soil cover, planting needs to be performed quickly after the first pass. Concomitantly, action is necessary to keep the existing straw cover, thus reducing the amendment effect on the soil and desertification. As deposition equipment is very expensive, there is a tendency to postpone deposition until just before planting to avoid possible losses. In this case, it is necessary to keep the straw during this period as well.

Cutting or topping methods and weed control affect straw decrease and retention, as well as future weed growth. Very low material discharge devices or devices needing the whole straw length as a permanent cover shall be adequate. Straw spreads within the seeding area shall be maintained, avoiding further labor. Devices to accelerate the mulching process from other materials are interesting. Straw spreading is essential to ensure weed control efficiency in reduced-row areas. Sowings besides contour intervals shall have direct seeding performed with a 90° angle between them and be maintained until primary weed control with its full cover.

5.3. Specific Techniques for Different Soils and Climates

To build organic matter in easily cultivatable and non-tropical soils, no-till principles can be used, but are not as crucial as they are in tropical soils. Very beneficial regenerative plant species like pig beans and pigeon peas can be seeded by aviation on large surfaces. Whereas pig-bean grows only in hot and dry climates, pigeon pea also grows in subtropical and tropical zones.

In hot climates, sod seeding is a method that specifically sows seeds into temporary or permanent coverage crops. In tropical and dry subtropical and internally culinary humid areas, horses have been used to till the soil with drill techniques to prepare them for planting.

In the mountains, a growth-reinforced combination of a small drill and rotorator nearly tills the soil. This is an organic and gentle method to prepare the soil for plant improvement. A direct mountain seeding method can also be used for climates that experience long terms of slow plant growth, often several years until ground coverage is sustainable.

On resources with high water retention, which have already been self-seeded by pioneer plants like alder or acacia, they rot fast, allowing seeding rows with the direct seeding technique. In regions with little rain, large seeds are being sowed at large distances from each other, whereas plants of crown-rose type are a soil insulator to shade the earth.

5. 3.1. Adapting No-Till Methods to Various Climate Zones

Crop rotation becomes even more essential to no-till in cold climates, which produce more weeds and insect pests through shorter growing seasons. A cover crop mix (which includes many species) will out-compete these pests, and crop rotation will lead to less of them. Crop rotation also enhances the self-organizing pattern of plant community development (including biodiversity) that covers crop mixtures. In wetter, equatorial sub-humid climates, on the other hand, with their potentially greater rainfall and fewer soil-building freezing-thawing cycles, compaction-busting tap-rooted species must be part of the mix and there is certainly no need to learn how to increase your rainfall, so great advice for no-till soil moisture management in humid climates is to never let your soil dry out.

There is no single answer to the question of the amount of biomass to lay down on the soil at the time of cash crop sowing, and the timing of that cash crop sowing, but as long as you begin in a way that the mix can quickly self-organize into an optimal rate of green and brown, it is easy to manage this issue properly, regardless of your climate. I recommend herbaceous meadow-plant diversity in the mix, in contrast to seed-harboring diversity.

Chapter 6. Introduction to Composting

Composting is the fundamental nutrient-dense produce for forest and crop garden applications for fertility and soil quality. It is the master concept of living, nutrient-dense soil which is the lifeblood, and optimum fertility can be achieved whatever soil type, pH, macro, and micronutrients to meet nutrient quality in the maturity of designed composts by the design of nature. All specifically produced composts are based on immature consulting compositions and mature consulting compost. They are unique to benefic ally populations that maintain the health and well-being of plants for the producer and the consumer, derived from compatible materials that decompose in nature's specified period, honor the designed needs of essential plant nutrients in mature structures, maintain their specific nature and are affordable and available.

The composting process is the act of composting which is always performed even when humans are not involved. Most living organisms have a defined lifespan, and when life comes to an end, the remnant material will compost or decompose over time. The act of composting can be broken down into several defined processes of this life and death-cycle. Composting can be divided into three critical phases: initiating, self-assembling, and the arrival of the final significant populations of consistent beneficial allies. Each phase within itself also provides detailed turning points that alleviate the concerns of odor, pests, and volumes safely and effectively placed active populations. Composting by design for the users can bypass the act of placing good plant nutrients within regeneration crops and replace them with multidimensional measures that superimpose new species and bio and oligo nutrient extracts. Our respective command of the needs of these bio, micro, and oligo nutrients will establish the laboratory within ourselves to comprehend how quickly these composting "design" modifications can occur.

6.1. The Science and Benefits of Composting

Composting is a natural process of decomposition. Through this process, organic waste is converted into compost, which is a valuable material for gardening and farming, enabling a good soil structure that retains moisture and nutrients. Composting is one of the most highly effective waste treatment and recycling techniques. We can put to good use the yard, food, and other organic wastes that we generate by composting them. Organic waste recycling techniques such as composting can significantly reduce waste disposal costs.

In addition to recycling organic wastes and providing a valuable soil amendment, composting provides a way to reduce organic waste at landfills. It reuses yard, food, and other organic waste on-site. It increases the soil water-holding capacity and soil aeration. It provides a slow-release form of nutrients, reduces erosion, aids in the destabilization of soil hydrogen to enhance percolation, and can be used to biologically fight plant diseases when applied to soils. It can reduce the need for pesticides, concentrate and eliminate pollutants, and can be used to clean hazardous waste sites. It is a choice for meeting water quality standards and is a good donor for wetlands or habitat restoration. It will help meet the criteria of urban runoff control and reduce the need for commercial rock fertilizers. It provides nutrients to trees, shrubs, flowers, lawns, and valuable plants, and enhances crop yields. Composting is prolific. It does not contribute to global warming – composting is a carbon sink. Also, sustained yields are supported.

There are many methods and theories for composting. Some of them are more involved and are more applicable at commercial scales, and some are basic and doable at the domestic level by homeowners. What is most important about composting is that you must live regularly! Only by experimenting and finding results with techniques, often centuries old, can you increase your understanding of the fascinating world of soil ecology dominant in composting.

6.1.1. Types of Composting Systems

Composting is not a new concept, even in developed countries, as they have been doing it for generations. Composting farm waste is also done to avoid the cost of materials such as fym, straw, and leaves for soil protection. To achieve higher efficiency and manageability of the composting process, different types of composting bins and boxes are built. From small wooden or plastic bins to huge thermal composters with forced aeration and heat recovery, almost anything is possible, depending on your budget. Many systems are sold as "patent" inventions, but they are often just recognizable designs and do not deserve the investment. Keep in mind the established costs per produced compost.

Different types of systems require different efforts to construct, such as material chopping and turning, disturbance intensity at the existing habitat site, compost stability, resource training content, as well as the quality of the final product and the costs of necessary additional process support and area requirements. There are five different levels of cost-effectiveness, recognizing differing physical quality and working inputs for small-size (plastic) bins and large open-air windrow and pile systems. Based on the location and the existence of specific pre-process treatment, such as the type of cutting and sifting tools needed, the potential system designs proposed fall between these two relevant extreme types. For more mobile and manual-intensive features, if one goes horizontally along "low" landfill-oriented technologies, free-standing vertical units come and slot in between aerobic windrows and horizontal worm composters as "economically optimized" urban composting equipment. Keep in mind that the type and quality of the compost do not solely depend on the construction of the composting device around your waste but also rely considerably on the feedstock preparation and management running in the house before you make any material choices going to your outdoor box.

Types of Composting Systems

1. Small Wooden or Plastic Bins

Compact and easy to manage, these bins are ideal for small-scale composting in urban or small gardens. it is typically made from durable plastic or wood, these bins can be purchased or built at home. it has Low cost and effort and is suitable for households with limited space and composting needs. Produces good-quality compost if managed well.

2. Thermal Composters with Forced Aeration and Heat Recovery

It has advanced systems designed for higher efficiency by accelerating the composting process through aeration and heat management. Involves more sophisticated technology, often requiring a higher initial investment. It has high cost and effort;

best for those with larger budgets and a need for rapid composting.

Produces high-quality compost quickly.

3. Open-Air Windrow and Pile Systems

Large, open piles or rows of compost material are turned regularly to maintain aeration. Simple and cost-effective, suitable for large volumes of compostable material. Moderate to high effort; requires space and regular turning of the compost. Good-quality compost, though it requires more time and effort.

4. Free-Standing Vertical Units

Vertical composting units save space and often incorporate features from both aerobic and worm composting systems. Moderately complex; can be purchased or built with more investment in materials. Medium cost and effort; suitable for urban environments where space is a premium. Produces efficient and high-quality compost with proper management.

5. Horizontal Worm Composters

This system uses worms to break down organic material, creating nutrient-rich vermicompost. It can be built using simple materials or purchased as ready-made systems. It has low to moderate cost and effort; and requires maintenance of worm habitat. Produces high-quality compost with minimal odor and space requirements.

Consider the following while choosing a Composting System

1. **Budget:** Determine how much you are willing to invest in your composting system. While some systems require significant upfront costs, they might offer long-term benefits in efficiency and compost quality.

2. **Space:** Assess the available space for composting. Larger systems like windrows need more area, while vertical units are suitable for smaller spaces.

3. **Effort and Time:** Consider the amount of time and effort you can dedicate to maintaining the

compost. Some systems require regular turning and monitoring, while others are more hands-off.

4. **Feedstock Preparation**: The type and quality of the compost also depend on how you prepare the organic material before it goes into the composting system. Chopping and sifting tools may be necessary for some systems to enhance efficiency and compost quality.

5. **Environmental Impact:** Consider the environmental impact of your composting system. Systems with forced aeration and heat recovery might be more efficient but can have a higher ecological footprint due to the technology used.

6.2. Composting Materials

Green materials (risers) or carbon-based materials are usually mixed with wet (nitro) materials. Microbes (nitro) break these organic materials into substances that can be absorbed by plants more readily than refining them by other means. When the ratio of green materials to dry substances (risers) is optimized, the compost's working purpose may also be maximized.

Green Contents

Each "green" material should be stirred and dropped through the container. Cut or strip the large slices to a specified size. Add green materials up to a depth of 3 cm. Add a layer of coarse materials like coffee blends or twigs at this degree and sprinkle some soil in the layer, but do not add any bone meal before constructing another string of composted material, adding bone meal, and stirring the leaves compost. Make your caramel water. Then add moisture to your compost, about 50 percent of water saturation, in the middle of the thumb and the finger stick together but not be safe.

Large Chip Size

While chopping, use a mid-sized stalk or hedge of the cutting device. Change. The worms remain within the compost bin. To enhance airflow during compost, we need to change the surface of the crude carbon ratio to no less than 16 carbon-hydrate air after continuing to stir the compost. After you have improved the appearance, you want to add the compost when you see no more heat in the compost middle. However, your compost is still wet and missed the water for 45 minutes. This helps to avoid washing and makes most of the nitrogen materials run out of water.

6.2.1. Carbon-Rich Materials

It is important to return carbon to depleted lands because soils rich in carbon sustain a rich landscape, with grasslands being one of the most carbon-dense of land ecosystems. When grasslands are tilled, many centuries of photosynthesis oligomerized into carbon compounds are decomposed and released into the atmosphere by microbes. This decomposition weakens the soils to stormwater and biochar erosion and diminishes biological and insect glider diversity to the point that grasses can no longer thrive. This is a competitive relationship created unknowingly by farmers and commercial agriculture.

Even when crop growth depletes the ground of carbon, regenerative farmers know that it is not their turn to profit from the brewers' sugar produced by their plants. Even mature plants have a complex but temporary use for it. But only mature prairies have evolved to store the excess carbon and to use the excessively available carbon as a unit of exchange, not only for funds. Roots were developed not centuries ago to repay rhizophagy and glomalinase by providing nutritious neurological carbon used by roots, aphids, and plants. Grasslands use the trade, building with roots and mosses for the middle class.

6.2.2. Nitrogen-Rich Materials

If you have been reading composting books, you may have come across the term "C/N" ratio. It refers to the carbon-nitrogen ratio in the compost material mix. For the most efficient and balanced composting, materials with a C/N ratio in the range of 15 parts carbon to 1 part nitrogen, to 30 parts carbon to 1 part nitrogen are necessary. Very little will break down if the C/N ratio is under 15:1 of carbon to nitrogen. When very high carbohydrate harvested for composting feedstocks are used, the C/N ratio can be from 10 to 20 parts of carbon to 1 part of nitrogen. Nitrogen is derived from the protein in nonwoody plants, seed meals, and non-polluted sewage sludge. Animal manure is a nitrogen source too. Blood and bone meal are listed as nitrogen sources in the compost recipes found in gardening books. Due to their animal source, these are not vegan compost starter materials. Plants with legumes (peas, beans, peanuts, clovers, members of the Mimosaceae, Fabaceae, or Papilionaceae families, etc.) form nitrogen-fixing root nodules due to being hosts for specific bacteria. The glandular swellings on the roots of the plant are nitrogen-rich modules which, when they are seventy-five percent brown, make great C/N ratio vegetable waste and other picadores for instant, vegan, nitrogen-rich compost starter.

6.2.3. Other Essential Ingredients

Microbial inoculants: In my experience, the first amendment to add is a microbe-rich inoculant. The desired final product for a microbial inoculum could be called Super High Brix Extract. It helps to feed the ensuing mineralization of soil minerals by externalizing, enhancing, and further introducing an abundance of beneficial organisms into the process. The immediate and ultimate awakening of life force from compost begins at the inception and continues through the incubation of life-propagating bacteria and fungi. It is the all-important life force that animates the biological materials in the compost heap, lest they remain moribund until eaten, soil digestion, and mineral release by thriving microbial life in the soil.

Carbonaceous matter: Providing crackling carbonaceous matter known as "Greens" is important for attaining a stable mature product and tempestuously stimulating nutrients by amplifying the capacity of the microbes and soil. Carbons are conduits that stimulate the absorption of macro- and micro-nutrients. Green vegetation, legume plants, animal food, manure, or silage can provide carbonaceous materials. Carbon rocks the soil microbiome, together with macro and micronutrients like rock dust, thus affecting enhanced soil, crop, and compost vitality. This is especially important at this beginning stage when what exists are life-forming morsels that will benefit from carbon's soothing presence.

6.3. Composting Process

The composting process involves the physical breakdown, chemical conversion, and microbial degradation of organic materials in the environment under controlled conditions. In the process, various organic compounds are oxidized, and the reduction of the organic C and N content of the material is achieved. The plant material plays a passive role in its transformation because the extensive enzymatic and metabolic activities of microorganisms are the likely causes of the sequential steps that occur during composting. The heat produced in this phase is usually enough to elevate the core body temperature of the mass of raw material to between 60°C and 70°C at its peak. Controlled heat, air, and moisture must be provided to the microorganisms in the decomposing mass to ensure that they act in a coordinated manner. Grower manipulation of these composting parameters will determine which microbial taxa become active to decompose the materials.

The physical activities of fermentative microorganisms, as well as the decomposition of complex organic compounds into intermediate metabolites such as organic acids, are the main cause of the temperature increases that occur in phase I. The beneficial result of this phase is the heating of the plant. Grazing and animal waste contribute nitrogen (N) in quantities that surpass the needs of the fermentative bacteria, creating an excess in the mixture. This results in high rates of N mineralization and ammonia volatilization, which will increase the pH, or create a high pH if there is an early failure. Most of the weeds are killed during this period, except for the seeds that can withstand high temperatures.

6.3.1. Aerobic vs. Anaerobic Decomposition

Aerobic decomposition is when organic materials are broken down by microbes in the presence of air. Anaerobic decomposition is the breakdown of organic materials by microbes without air. In nature, both aerobic and anaerobic decomposition play the crucial role of breaking down waste materials so that they can be used again by other living organisms. In our compost piles, we want the decomposition process to be aerobic.

In an aerobic environment, the primary result of decomposing a carbon material (straw, wood chips, or a cucumber, for example) is the production of carbon dioxide and other gases. These gases account for a significant portion of the weight loss observed during decomposition. The finished compost that is left behind will have a considerable volume reduction when compared to the bulky waste materials that went into the process.

Anaerobic decomposition smells bad and takes significantly longer than aerobic decomposition, and the end product is often runoff that contaminates ground and surface waters. The aerobic vs. anaerobic choice is often not very clear even when we are trying to make compost aerobically. To be successful we have to think about how to install good conditions for decomposition in our compost piles including creating an aerobic environment with lots of air in the piles.

The heat generated by the decomposing organic materials is often the most tangible sign of good decomposition. We say that the compost is "hot as a snot" which means that it has worked well and has a lot of heat energy in it. Compost that does not heat up in this way will become anaerobic and smelly because not enough air is getting into the pile. We must aerate it by turning the pile over. For both aerating the pile and providing good airflow, some kind of porosity has to be provided in the pile.

6.3.2. Temperature and Moisture Control

Control of temperature and moisture, and maintenance of good airflow, are the keys to successful composting. High temperatures are an integral part of the process that enables the compost to break down a wide range of compounds and pathogenic organisms. In contrast, little can happen to solid organic materials without the intervention of such heat. The conventional approach is, therefore, to ensure that composting always, or almost always, takes place at high temperatures. These optimum temperatures occur when the moisture content of the composting materials is also close to that needed to produce cooked brown rice. Moisture management, therefore, often involves careful maintenance of the initial structure of the composting material or the development of alternative approaches.

Proper temperature control is usually achieved by laying out the composting materials according to their initial moisture content, injecting some air to start the process, and then walking away for three or four days. The most common approach to achieving the initial moisture content needed for creating high, persistent temperatures is to maintain the moisture for a few weeks before composting, and then to add dry materials during composting until the right number is found. In other words, extra dry materials are added to the more negative water fraction (-1). This approach usually generates very high temperatures but is very expensive in terms of the extra effort needed to handle dry materials. Starting composting with more standard recipes generally results in higher moisture content, as microbial respiration produces water as a by-product.

6.4. Mastering Composting Techniques

System dynamics within composting are complex and can be rather overwhelming. With more and more households and communities embracing composting to create a circular nutrient cycle, it is equally important to understand the science behind good composting techniques. While we may throw anything into a compost bin or pile, then turn it once in a while and water it a bit, the rules are a bit more intricate than that. With good knowledge, we can achieve high-quality composting in less time. There are different paths to good composting, based on different types of organic material, such as carbon-based or nitrogen-based organic matter. All share certain common controls.

Capable tools can ease the workload. Common tools include simple plastic bins, bins using wood furniture made of untreated wood, to commercial tumbler types. Aerate and water, if necessary, to keep compost alive. To produce good-quality compost, aeration should be provided at regular intervals before the material self-compacts. Depending on the type of material used, moisture can be added back to the pile. The moisture of the compost material should be around 50-60% and will feel moist to hold but should not leave moisture on your hands when testing. While adding water, do not soak to the point of saturation. The heap should not be directly waterlogged, rainwater storage due to wet compost can lead to nutrient leaching. The quality of the finished compost is determined by the final decomposition performance at the end of the composting cycle through temperature, stability, and maturity.

6.4.1. Detailed Step-by-Step Composting Guide

The production of quality and abundant compost is significant if the emphasis of your agricultural business is regenerative agriculture. It is also super important if you intend to utilize no-till or zero-till practices. Many people start their regenerative agriculture journey with composting and, unfortunately, produce a low-quality product that feeds disease-cycle stimulation, oxidative stress in crops, and inferior resilience in crops to abiotic stress. The main reasons for creating such products are the adoption of traditional recipes, a shallow understanding of the reason for the recipe, and not using a beneficial enough inoculation. These recipes are suitable for small-scale PM dogmatic-era organic growing, not for regenerative farming at world-leading rates. Use solid science instead of adepts' 0.7% carbon padawan organic soil science.

Here is a brief version of concrete step-by-step composting instructions. The main point about composting is that it should be carried out with the thought of feeding beneficial soil microbiology.

1. Collect browns (high-carbon material, such as straw) and greens (high-nitrogen material, such as fresh grass).

2. Grind brown material with a shredder to no more than 3-inch long pieces.

3. Green chop the greens by finely cutting them with harvesting scissors.

4. Pre-soaking is very important because dry raw materials do not compost. Never pre-soak indoors due to the strong smell produced.

5. Start composting with specific inoculation and moisture. Focus on thermophilic alpha bacillus.

6. Put the compost on grass bales. It helps heat generation, air keeps exchanging, moisture keeps releasing, etc.

7. Evaluation points during composting: if compost does not heat up and also does not produce a fermented smell or usable products after fermented smell, it was over-matured compost.

6.4.2. Turning and Monitoring the Compost Pile

The least amount of turns, however, promotes the most prolific growth as broader regions and concentric stratifications of diverse organisms attempt to and then vie for their preferred choice of compost. Fungi prefer to be surrounded by fungi. Bacteria prefer to be surrounded by bacteria. Actinomycetes prefer to be surrounded by other actinomycetes. And so on. As much as we as humans might get along with almost anybody under most circumstances, compost communities of organisms don't seem to cross the road and befriend others that are defined by relatively few species of organisms in any meaningful placement towards other breeds of organisms. The competition is so fierce that they are at many times their peak quantities. They are not friendly to other dissimilar qualities of offerings provided by diverse biomes, and they hang out essentially with themselves.

The first turn should be when the pile has performed its first big decline in temperature. Turning up to 30 days, initially in cooler weather (or when it begins to get smelly) and close to 55°F, or 18°C, if no temperature probe is used, is generally the best standard. After turning the pile, it will begin to heat again as the temperature of the pile is raised. If there is still enough food and as yet no problems, the pile or windrow will be reheated to around 140 to 149°F, or 60-6°C, and then decline again. Then other turns may be required. These second/secondary heating turns will perk up the material. No more than three turns may be necessary or conducive for making the most nutrient-rich compost in alright. Every turn introduces unwanted competition. Every turn brings the temperature down. And every turn brings the opportunity for a big fungal pop to slide in and begin feasting big time.

6.5. Troubleshooting Composting Issues

In subsequent weeks of composting, avoid depleting the pile from moisture by overwatering with too high a nitrogen content. In arid patches, add a bit of water to the self-brewing pile. There should be plenty of useful woodsy stuff in ready supply, such as conifer needles, pinecones, twigs, shredded leaves, bark, etc., which can be added to the pile freely. Compost piles should also be covered with a tarp to shield them from the elements, but the material needs regular and repetitive turning to fulfill its needed purpose. If you have built your pile with regular yard waste, remember to balance the carbon over the scale of the nitrogen at a respective range of 7:11.

If the pile is too dry: Too little moisture can cause a compost pile to "freeze" up. Try watering the material as you layer it. If it gets too dry later on, consider covering the whole thing with a tarp for a bit before restarting the layering process.

If the pile is too wet: Genuinely soggy, over-wet material should be handled by removing some of the material and running it through a chipper shredder or spreading it out to increase the surface area available to the air. Avoid piling it higher than 1.5 meters. Also, shift the material by distributing the wettest parts of the pile uniformly, then turn the material and remix it to distribute the water, which helps to manage the moisture as evenly as possible.

If the pile smells: A pile that smells sour can be improved by turning it to distribute the oxygen. The smell is due to toxins produced when bacteria are waterlogged and deplete the oxygen. The solution is to slice or prick the material with a rod to unwind the air pockets and distribute the air. Odorous air can puff out as the damp air escapes. Continue to turn the material every other day to redistribute the air and avoid odor. If no unpleasant aroma develops, an active critter community can consume and transform the pile faster.

6.5.1. Identifying and Solving Common Problems

Ammonia-smelling compost or nasty slimy compost or stinky or white grubs in a heap

• **Problem identification:** You will soon begin to dislike having to check a sickly-smelling heap that must also release nasty gases like ammonia (highly engaging gas on smell because it involves urea!) This unhealthy brew will certainly repurpose you to wear more cotton and to work closer to the heap; both strategies can sharply heighten your awareness of self-care and the heap's preferences! This exercise should reassure you that you are correctly accusing excess nitrogen of causing the ammonia smell.

Insufficient or slow composting time

• **Problem identification:** Objectively evaluating compost is challenging for new composters but is a necessary skill. Feel, not smell, your compost on day 4, day 7, and again on day 14 to help build your knowledge base. Refer back to this test regularly to keep making better and better compost. If your composting process produced lots of sludge or partly finished compost, you evaluated incorrectly on one of these days.

• **Solution:** Cause and maturity diagnostics will usually confirm that the reason for slow composting is cold core temperatures, insufficient ventilation and/or mixing, compaction, or excessive rain. It may also be due to dry woody materials in the mix. The active phase of the heap should not exceed 20-21 days. Informed composters will schedule harvest accordingly, compressing and opening windows for seasonal demands.

Part III: Advanced Techniques

Despite the numerous simple techniques often discussed for water management on farmland, there are numerous ways we can use technology and innovation to better manage water, fully enabling regenerative agriculture on farms of all types.

The science of rainwater harvesting, or how water can most effectively be added to the ground to sustain crops, is a key aspect of regenerative agriculture. While larger earth-based strategies such as ponds and swales are often talked about in environmental discussions, this section will delve into the specifics of how rain or surface greywater can be most strategically added to the soil. Key problems are solved, such as off-grid solar-powered pumps pulling dirty water from the infrastructure attached to a greenhouse, and filtering out the compost, thereby supplying clean water before being sent through hose bibs or misters. This section presents several short-term and self-immediately available solutions for common water-related farming hassles; for example, the human body is composed largely of greywater, cleaned perfectly by the liver and kidneys before being excreted with both nutrients and other substances that can be beneficial to the plants.

Chapter 7. Efficient Water Management

Efficient water management is crucial for sustainable agriculture and environmental stewardship. As water resources become increasingly scarce and demand continues to rise, implementing effective water management strategies is essential to conserve water and enhance its availability and quality for agricultural use and ecosystem health.

Water conservation in agriculture involves a shift from merely limiting damage and depletion to actively replenishing groundwater and soil water bodies. This approach, known as regenerative water management, focuses on reversing the negative impacts of human over-extraction of water and soil degradation. Living soils with high organic matter content can hold more water, providing numerous benefits such as improved crop yields and enhanced soil health. Increasing the organic matter in soil enhances its water-holding capacity. Practices such as cover cropping, reduced tillage, and the addition of compost and organic fertilizers contribute to building soil organic matter. Introducing diverse tree species and perennial plants can improve water retention in the soil. These plants have deep root systems that help infiltrate water and reduce runoff. Planting trees on degraded lands helps restore ecosystems and improves the water cycle. Forests play a crucial role in water filtration and maintaining the hydrological balance.

Sustainable grazing practices, such as rotational grazing, help maintain ground cover and prevent soil erosion. This, in turn, enhances the soil's ability to retain water. Constructing ponds, reservoirs, and other water storage systems can capture and store rainwater for use during dry periods. These structures also help in recharging groundwater. Techniques like contour farming, terracing, and the use of swales can reduce runoff and encourage water infiltration into the soil.

Regenerative water management techniques mimic natural processes to enhance the ecological functions of water. These techniques not only improve agricultural productivity but also contribute to the resilience and sustainability of ecosystems. Creating hydrated mineral matrices in soils helps in organizing water and improving its availability to plants. This involves using soil amendments that increase the soil's capacity to hold and distribute water efficiently. Effervescent outgassing, which involves the release of gases from the soil, can improve soil structure and water retention. It also contributes to the formation of stable carbon compounds that enhance soil health and water-holding capacity.

Promoting the growth of beneficial soil microorganisms can enhance soil fertility and water management. These microorganisms help in decomposing organic matter, releasing nutrients, and improving soil structure. Designing landscapes to mimic natural water accumulation processes can help restore the hydrological cycle. This includes creating wetlands, riparian buffers, and other natural water storage systems. Infiltration basins, which are designed to capture runoff and allow it to infiltrate into the ground, replenish groundwater supplies and are particularly useful in areas with heavy rainfall and high runoff. Collecting and storing rainwater from roofs and other surfaces can provide a reliable water source for agricultural and domestic use, reducing dependence on groundwater and surface water sources.

Real-world examples of successful water management provide valuable insights into the practical application of these techniques. In Argentina, farmers have developed specific techniques suited to the country's dry climate. These include contour plowing, which reduces erosion and improves water infiltration; terracing, which slows down water flow and increases water absorption by the soil; and dryland farming, which utilizes drought-resistant crop varieties and planting techniques that conserve soil moisture. These methods have shown significant increases in water use efficiency and crop yields, demonstrating the potential of regenerative water management in arid regions.

In the United States, farmers practicing biologically oriented agriculture have adopted various water management techniques, including cover cropping to protect the soil, enhance organic matter, and improve water retention; no-till farming to reduce soil disturbance and maintain soil structure and moisture; and integrated water management systems that combine rainwater harvesting, irrigation efficiency improvements, and the use of soil moisture sensors to optimize water use. These practices have led to improved water use efficiency, higher crop productivity, and better soil health. The results indicate that regenerative water management can be highly effective across different climatic conditions.

Efficient water management is vital for sustainable agriculture and environmental health. By adopting regenerative water management strategies and techniques, we can enhance water conservation, improve soil health, and build resilient ecosystems. The success of these approaches in various regions underscores their potential for widespread application. As we continue to refine and implement these practices, we move closer to achieving a sustainable and water-secure future.

Chapter 8. Integrated Pest Management

The dependence on synthetic chemical pesticides is a significant shortcoming of most cropping systems at present. This includes conventional agriculture, but even more so modern organic systems. In the global south, toxin products like endosulfan compounds or other highly toxic substances are widely used because they are locally available or very cheap. The consequence is that residues can be found on the fruits or vegetables or the users suffer an enormous burden due to intoxication and poisoning. The destruction of the food-chain enemies of the pests and crop diseases, which are beneficial organisms for example solitary bees, insects and wild plants, is another related environmental problem. So it is time to look for alternative goals like "pest management in balance" or sustainable integrated pest management (IPM) for the future.

The knowledge about beneficial organisms and interactions is still often very low so new pest management strategies always have to be worked out on a local basis. As the measures described abuse encouraged beneficial organisms either already found or established on farms or waste or wild herbaceous vegetation at the edge or in small strips within the agroecosystem are supported. It is interesting because the prevention is relatively simple, safe (risk-free) and cheap. Sowing of vegetation has only to be done for a short time; most of the species are self-perpetuating and reinstate their populations automatically. Therefore insurance policies that support local uses are accepted by many.

8.1. Natural Pest Control Methods

There are several ways to change the natural cycle to help ensure that the crop can stay healthy. One way is to avoid planting species of crops that these pests cannot subsist on. Planting a small field of another type of crop that will be set aside before or during the growth of the main crop, or only growing a single crop with some land being left untilled and fallow, may confuse the pests to the point where they won't find the crop or nutrient condition to overwinter as a cycle. Combine the nutritional information with a seed or seedling that will confuse the insect at a time when it is most dangerous, and the effect will be strong.

It is rare for an insect to kill a crop. Even the most effective pests, if not killed by natural predators or modified weather changes, can only survive on the field crops as long as they remain there. They will eat it down to the ground and start dying off from the top, as bacteria start breaking down the first plant to die and sugar helps to develop the bacteria population. Predators can help to either prevent a cycle by eating them or prevent the growth of numbers until the crop is beyond a stage where the insect can cause much economic loss. Predators, mostly birds, and other animals, can make the few insects more valuable by eating them, and by eating seeds, eating dead plants, and such. In this way, the potential pest pattern is kept in check.

8.2. Creating a Balanced, Pest-Resistant Environment

How satisfied are you with your current pest control program? For many farmers, signals such as crop damage, a sudden surge in pest numbers, and/or a spate of disappointing harvests have encouraged a reliance on non-compatible chemicals. You can increase the profitability of your operation by changing your thinking. Nutrients may be pushed into the plant but can be quickly lost, not used by the plant, leach into the groundwater, or be made into pollutants secondary to other growth problems in the plant. Provide for the soil food web. The food web provides nutrients to the plant in exactly the form and amount the plant needs to grow its best. No plant diseases, no pests. These are the first Principles of Soil Food Web Management.

Although many so-called pests in our cropping systems are indeed native organisms that have functions when the percentage of populations of organisms that have a particular survival strategy is near 0, other pests are also likely to be native organisms that have switched the survival strategy they employ in response to the farmer's use of chemicals, fertilizers, improper tillage, and/or variety selection that does not provide for all the organisms in the surroundings. We are simply creating an unbalanced environment and then having to apply other chemicals to make up for that imbalance we caused. Re-examine these inputs, choose to produce food for the soil food web, and you will have a balanced, pest-resistant, and disease-suppressive growing environment.

8.3. Addressing Common Pest Challenges

Pest management is about designing an agroecological system that provides efficient use of resources to nurture strong, diverse communities of soil organisms while providing minimal resources to pests. Polyculture, well-designed crop rotations, hedgerows, soil health, and the skillful use of cover crops and animal impact are typical tools in pest management in regenerative systems, as are soil inoculation products, biostimulants, insectary plantings, and beneficial nematodes and insects, among others.

Genetic resistance to pest challenges is also emphasized, and a broad definition of resistance includes plant secondary compounds, increased root exudation of compounds that promote beneficial microorganisms, plant root traits that enhance nutrient acquisition, leaves that retain water, increased reproductive fitness (i.e., seed set), and a higher root: shoot ratio that reduces the flow of nutrients from the plant to the rhizosphere.

Pests tend to be most severe when monocultures and other imbalances in crop communities and associated soil organism communities occur. In agriculture, chemicals designed to get rid of pests disrupt ecosystems and microbiomes, providing favorable conditions for pest populations to explode. Instead, in a regenerative system, the agroecologist learns how to attract or introduce natural balances of organisms that feed on pests or interact with plants and the soil to deter or physically block pest-related damage. Regression analysis and multivariate techniques can be used to relate pest damage levels to soil community/composition/function data, to identify key pest management opportunities in each field of interest. An ongoing "pitch and stitch" process of using multiple, small, natural, and synthetic discoveries to mitigate the impact of pest problems is a common topic in regenerative research programs.

Part IV: Practical Applications

There are countless species and individuals in a gram of soil, with contributions from many taxonomic groups (bacteria, fungi, micro-, and meso-fauna) spread across a range of soil habitats, including spatially-formed pore networks and intimately associated with root surfaces. As a result, life in the soil can improve the soil network and contribute to reduced runoff and erosion, which can enhance soil health. Whether or not the problem is realized depends on a variety of factors. This chapter outlines many of these issues and provides both evidence for and examples of ways to employ regenerative practices that can help to grow a greater soil diversity beneficial to both agriculture and the surrounding ecosystem. Moreover, it highlights the need for more research to better understand the mechanisms governing soil biological diversity to make more informed best practices.

Chapter 9. Soil Testing and Analysis

In setting priorities for a soil management program, a farmer may decide to alter any one of the six soil-forming factors, and all will greatly influence management decisions. In soil testing programs for both conventional unplowed and minimum-tilled fields, the traditional soil-agricultural management system is modified to some extent by local agronomic, meteorological, social, and economic conditions. Bear in mind, concerning any soil testing program, that it is a general guide and not a production recipe completely specific to your fields. Only good farming judgment will get the proper response from a typical soil testing program, from which treatment rates will be computed by the chemist concerning the metabolic cycles of plants and soil organisms. Only good farming judgment will get the proper response from a typical soil testing program.

This book details soil transformations and functions that occur within well-managed and natural agroecosystems. The bulk of it is devoted to the description of the properties of soil organic matter, several intriguing aspects of microbial physiology, and the intersection of many important soil transformations with that of the behavior of carbon in soils. The book simply educates the reader about how soils function through time and the impact of soil management on those functions. It is intended to begin the discussion of new agricultural research programs that will develop innovative strategies for diminishing the gap between actual and potential crop yield.

9.1. Essential Soil Tests and How to Conduct Them

Early soil testing is important to understand what is happening in your soil. While this is especially important as you learn how to manage your soil more effectively, regular testing is recommended so you do not create other problems while addressing those you originally identified. While soil tests vary greatly in costs, there is also great value in doing some simple home tests. As they are free, readily available and can be done on soil samples from a wide range of properties, they can be done cost-effectively and quite frequently. Needs and ability to pay for local laboratory rainfed and irrigated soil tests depend on available resources. Costs include sampling, analysis, and possibly extra advice. Basic advice from relatively simple tests done at home and discussed with neighbors can also be very useful and low cost.

Nutrient availability is a complex interaction of many factors including soil reaction (pH), soil mineral components, soil humus content, and biological components that can help release nutrients from organic matter. Many tests are available, including simple home tests that are often good at predicting nutrient uptake under particular farm conditions. In regenerative farming practices, the cheapest and easiest home test involves simply mixing equal quantities of soil and white vinegar (acid 5% acetic acid). If the soil sizzles strongly, it is likely to have very active nutrient release. Do this test on the soil horizon among plant roots. If a significant amount of bubbling occurs after 15 to 30 minutes as the vinegar dissolves, the soil should be good for plants. Usually, the quicker and stronger the sparkling, the greater the variation of available nutrients. The easiest specific soil nutrient tests identify N, P, and K.

9.2. Interpreting Soil Test Results for Actionable Insights

Soil test results can provide a wealth of insight into the biological, physical, and chemical properties in soils. These tests can be divided into two broad groups: biological assessments that help to understand the condition of the soil biological community and physical and chemical assessments that are typically referred to as soil tests.

Soil biological assessments generally revolve around optimizing the number and activity of particular groups such as the soil food web, focused on the interactions within and between minimal cultivation systems and the soil ecosystem. The physical tests are primarily focused on measuring soil structure and should consider a range of options to evaluate factors such as soil friability, moisture levels, waterproofing tendencies, pull, inverting or compacted layers, topsoil depth or $CaCO3$ percentage.

In order to budget and plan crop inputs and cultivations, it is important that information from the laboratory can be sincerely interpreted. The key to interpreting a suite of laboratory soil tests lies in understanding the soil processes being evaluated and the relationship these processes have with plant nutrition and the physical structure of the soil. The support links should be made from available test results back to the key soil indicators that we have interpreted from conducting our practical soil review. If the support links required to benchmark interpretations are successfully established, then solution options to sustainable cultivations will be more effective. The definitive outcomes will be evident when crop inputs are modified to match a more favorable soil environment.

BONUS

Instructions: Step-by-step guide on How to Collect Soil Samples

1. **Gather Supplies**: Obtain a soil probe or shovel, a clean plastic bucket, and plastic bags or containers.

2. **Select Sampling Locations**: Identify different areas of your field or garden to sample, ensuring they represent the overall soil conditions.

3. **Collect Samples**:
   - Remove any surface debris (leaves, rocks).
   - Insert the soil probe or dig with a shovel to a depth of 6-8 inches.
   - Collect a small amount of soil and place it in the bucket.

- Repeat at 5-10 locations within each area, mixing the soil in the bucket.

4. **Mix and Subsample**: Thoroughly mix the soil in the bucket and take a subsample of about 1-2 cups. Place this subsample in a plastic bag or container.

5. **Label Samples**: Label each bag with the sample location and date.

6. **Send for Testing**: Send the samples to a reputable soil testing lab following their submission guidelines.

Data Entry: Recording Soil Sample Information

File name	Description	Entries
Sample Location	Describe the area where the sample was collected	[e.g., North Field, Garden Bed 3]
Sample date	Date the sample was collected	[e.g., June 15, 2024]
Soil type	Description of the soil (if known)	[e.g., Sandy Loam]
pH level	Test for soil pH	[e.g., 6.5]
Nutrient levels	Test results for key nutrients	
- Nitrogen (N)	N content in ppm or mg/kg	[e.g., 20 ppm]
- Phosphorus (P)	P content in ppm or mg/kg	[e.g., 15 ppm]
- Potassium (K)	K content in ppm or mg/kg	[e.g., 120 ppm]
Organic matter	Percentage of organic matter in the soil	[e.g., 4%]

Other observations	Any additional notes or observations	[e.g., good earthworm activity]

Analysis: Interpreting Test Results and Planning Amendments

pH Level Interpretation	The ideal pH range for most crops is 6.0-7.0. If the pH is too low (acidic), consider adding lime. If too high (alkaline), consider adding sulfur or organic matter to lower the pH.
Nitrogen (N) Level Interpretation	Low N levels may indicate a need for nitrogen-rich amendments like compost or manure. High N levels suggest reducing N fertilizer application.
Phosphorus (P) Level Interpretation	Low P levels can be improved with phosphate fertilizers or bone meal. High P levels may require reducing P inputs to prevent environmental runoff.
Potassium (K) Level Interpretation	Low K levels can be corrected with potassium fertilizers like potash. High K levels usually do not pose a problem but monitor for potential imbalances with other nutrients.
Organic Matter Interpretation	Higher organic matter improves soil structure and fertility. If levels are low, add organic matter such as compost, cover crops, or mulch.

Other Observations and Actions	Note any other significant findings (e.g., compaction, poor drainage) and consider appropriate actions such as aeration, improving drainage, or planting cover crops.
Recommended Amendments and Practices	Based on the above interpretations, list specific amendments (e.g., lime, compost) and practices (e.g., cover cropping, crop rotation) to improve soil health and fertility.